591
CAR

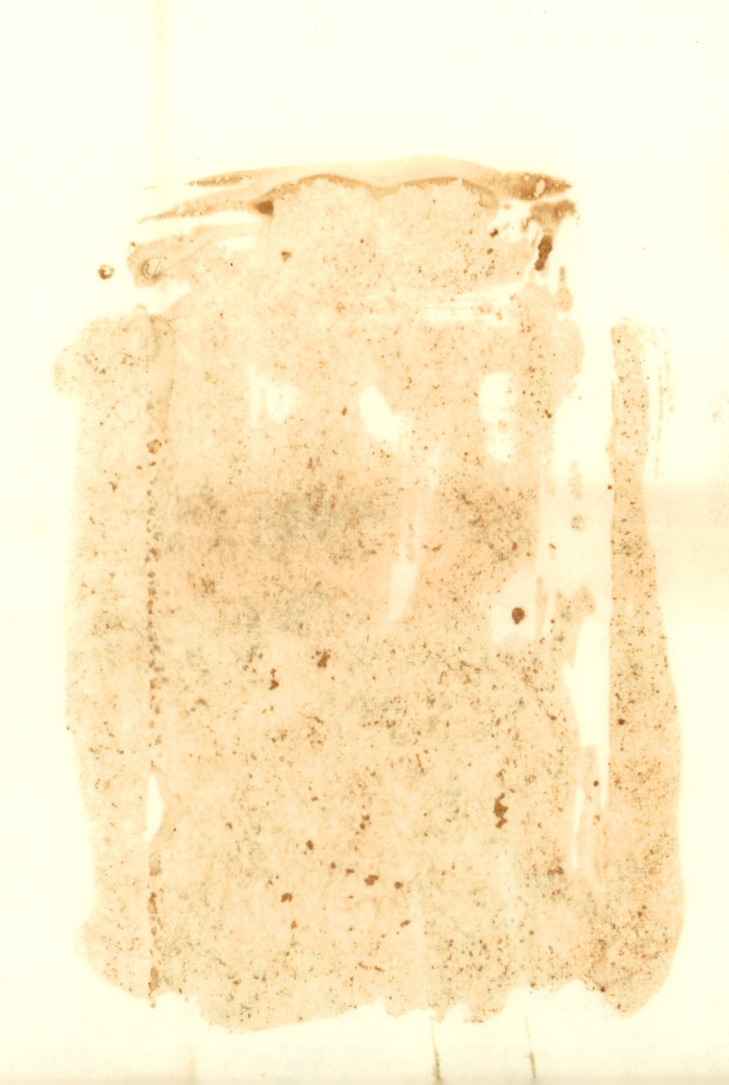

# THE BIZARRE ANIMALS

by
ROGER CARAS
foreword by
Roger Tory
Peterson

Barre/Westover Books, Inc., New York
Distributed by Crown Publishers, Inc.

DEDICATION . . . . .

This little book is for
Richard, David, Victoria, Geoffrey, and Susan
—in the order of their birth—
who may at times
find their uncle a tad bizarre.

© 1974 by Roger Caras

All rights reserved. No part of this book
may be reproduced or utilized in any form
or by any means, electronic or mechanical,
including photocopying, recording, or by
any information storage and retrieval
system, without permission in writing from
the publisher. Inquiries should be addressed
to Crown Publishers, Inc., 419 Park
Avenue South, New York, N. Y. 10016.

Library of Congress Catalog Card Number: 73-79125

Printed in the United States of America
Published simultaneously in Canada by General Publishing Company Limited

Prepared in cooperation with
Photo Researchers, Inc.
New York, New York.

Book design by L. A. Strickland

# INTRODUCTION

One dictionary gives the definition of *bizarre* as "markedly unusual in appearance, style, or general character; whimsically strange; odd" while another suggests "strikingly unconventional and far fetched in style or appearance; odd, grotesque." *Fantastic* is proposed as a synonym.

The origin of the word is not clear. *Bizarr* and *bizarro* can be found in old French and Spanish and variously is taken to mean brave and even handsome. In Basque, *bizar* means beard. If you add a masculine suffix, *-o,* you get he-man or spirited, in Basque, of course. As you look at the animals on the pages that follow keep these probable origins in mind.

The brave part of the traditional origin of bizarre might refer to creatures that *dare* to look different. The concept of *spirited* could be close to that, as well. The rest of the relationship is more obscure.

Indeed these are animals that are different in appearance. The idea that they *dare* look as they do, of course, is even more whimsical than their appearance. They are what they are because it is what they have to be in order *to be*. It is really quite as simple as that. Through the incredible maze of evolution, through the eternal imperative of trial and error they have arrived at a present state that balances the animal with the forces that conspire against it. No intelligence or plan was involved (unless you wish to acknowledge the Divine), at least none on the part of the creatures themselves or the billions, the trillions of ancestors that died testing out the approaches to the present condition. That is remarkable, when you come to think of it, truly as remarkable as the creatures themselves.

                                        Roger Caras
                                        East Hampton, N.Y.

# FOREWORD

Our species, *Homo sapiens,* is the dominant animal, the undisputed kingpin of a world in which systematists have catalogued more than 1,000,000 other species, from organisms as minute as protozoa to animals as ponderous as elephants and whales. The evolutionary process has given us birds as disparate as penguins and peacocks, mammals as preposterous as the platypus and the narwhal, insects as divergent as morpho butterflies and walkingsticks.

Among this diversity we find forms so bizarre that we ponder the reasons for such unlikely adornment. What survival advantage does the peacock gain from its over-sized tail with its intricate ocular pattern? Actually, the peacock has evolved into the gorgeous bird that it is because of the selective eye of the relatively demure hen. This is also true of pheasants, grouse, birds of paradise, and many other "arena" birds, the males of which strut, dance or display competitively. The females select the dandy of the lot and his special genes are then carried down the line.

Not all adornment is for sexual aggression or attraction. It may be camouflage as in the case of the stonefish that looks like a mossy rock, or the caterpillar that mimics a bird dropping or a twig. Bright color may be a garish warning of danger as in the coral snake, or it may signal inedibility as in some butterflies. Or pattern may frighten (as by means of huge false eyespots—a recurrent theme in nature).

Undeniably, the most bizarre forms of all are to be found in a drop of water or in the depths of the sea, shapes and structures completely alien to those on land. But whether or not we search for the fundamental reasons behind form, color and adornment, we cannot deny their infinite variety.

Roger Tory Peterson<br>
Old Lyme, Connecticut

## Cicada

There are several superlatives to describe the cicadas. They are the longest lived of all insects, some for as long as seventeen years. They are the loudest singers in the insect world, at least the males are as they stretch and relax corrugated drum membranes on their sides and call to their prospective mates. Despite their common names, "seven year locusts" etc., they are not locusts. In a way their lives are as bizarre as their appearance for most of it is spent underground. One species spends seventeen years crawling around in the underground gloom to emerge for only a few brief weeks of sunshine before dying.

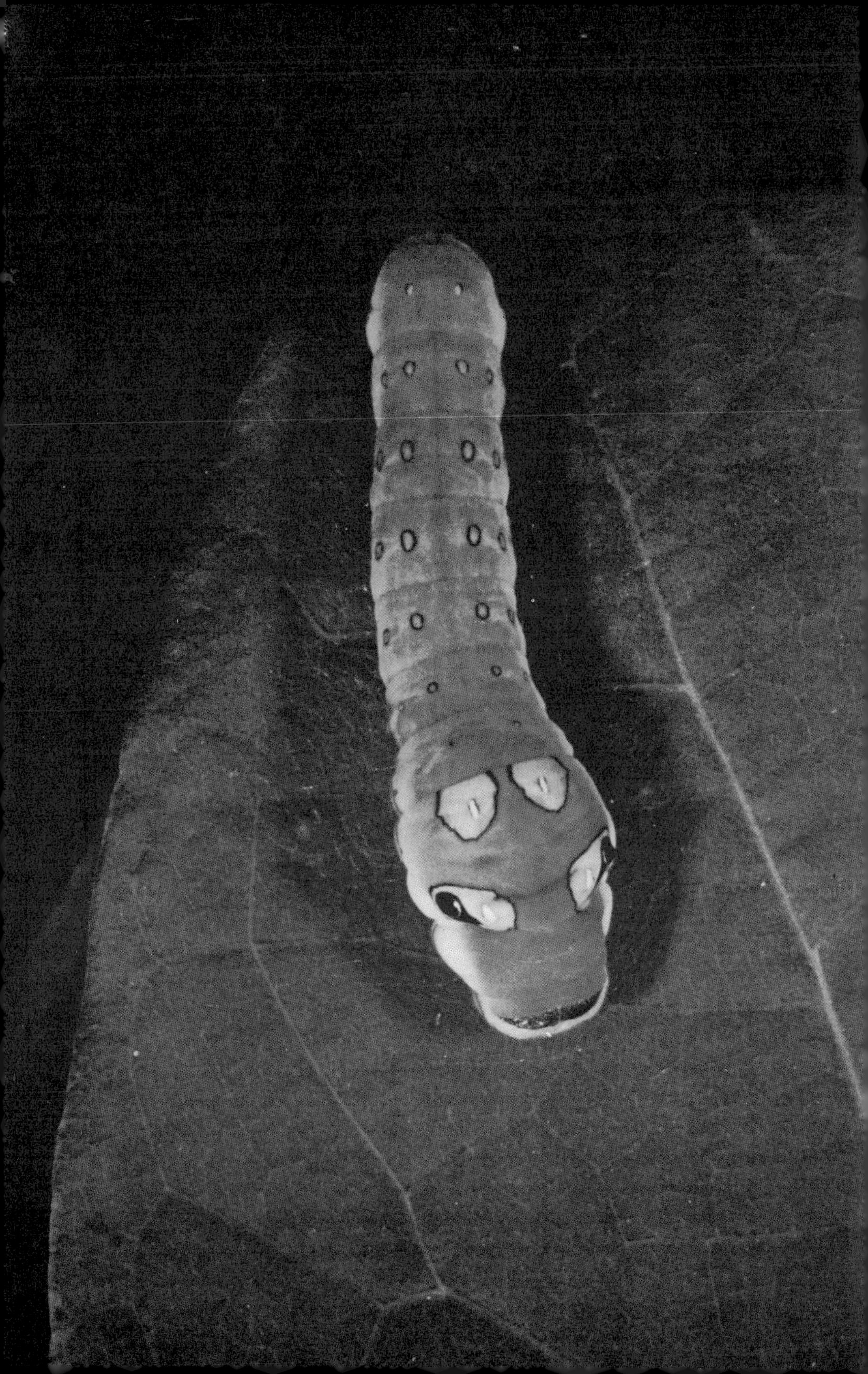

## Rhinoceros Beetle

These incredible looking beetles are successful, diverse animals that cope very well with a variety of environments. We may not understand all of the ramifications of their design but we can be sure the best way for this rhinoceros beetle to look is just the way it looks here. It can conceal itself better, it can intimidate the appropriate enemies more effectively and it can manipulate the things it must manipulate more efficiently because it appears to us the way it does. It can be hard to keep in mind that aesthetics just don't enter into the logic of natural design. It is all survival.

## Spicebush Butterfly Larva

Eyespots have nothing to do with the animal's real eyes. They are seen many, many times in the animal kingdom. In the case of this larval butterfly they give the impression that the animal is much larger than it really is. It looks more like a snake than a caterpillar when seen head on. That is their purpose—to discourage predators. In other species eyespots direct the attack of enemies away from vital organs and increase the animal's chances for escape.

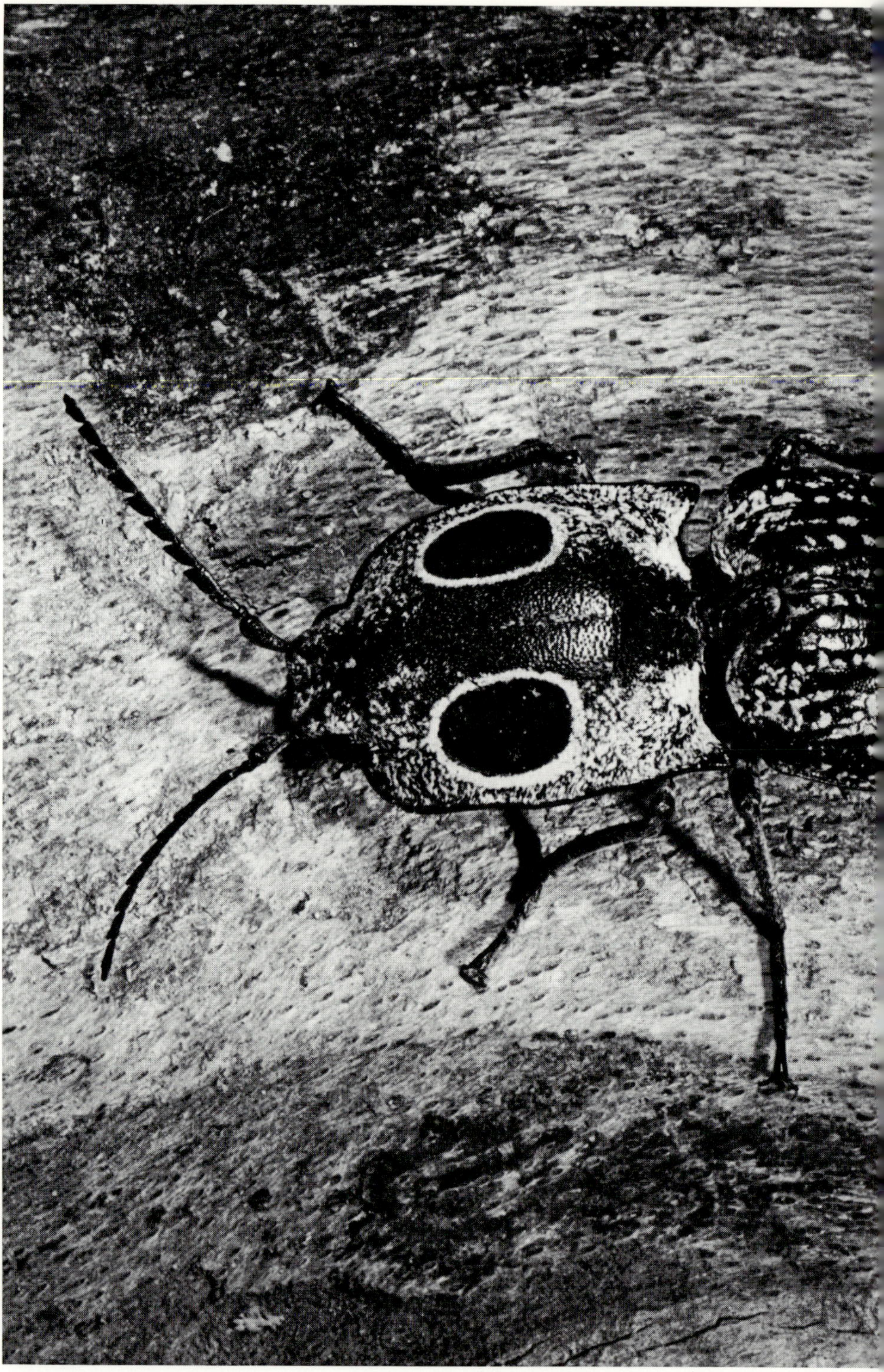

## Eyed Elator Beetle

Again we see eyespots. It is remarkable to think how they came to be. Once upon a time a bug of this or an ancestral species had markings that tended to confuse at least one kind of enemy. It was probably smaller and less distinct a mark than we see here. But, that mark tended to allow the animal that had it a little more life and a better chance to survive and breed. The mark was thus emphasized until it emerged as we see it now. It is a clear survival device. It will scare away some animals or at least hold them at bay long enough for the bug to escape and it will confuse others. Eyespots are remarkable products of evolution.

## Stag Beetle

The B.E.M.—the bug-eyed monster—of science fiction runs a poor second to our own world when we see it enlarged. This stag beetle is neither handsome nor ugly, it is functional. It may be startling in our view, as we see it rather larger than life, but to other beetles it is just fine. That is not because of its appearance, but because of its chemistry. It smells right at the right time and it may make the right sounds and movements. Smell, posture, sound—these are the things that make insects attractive to other insects.

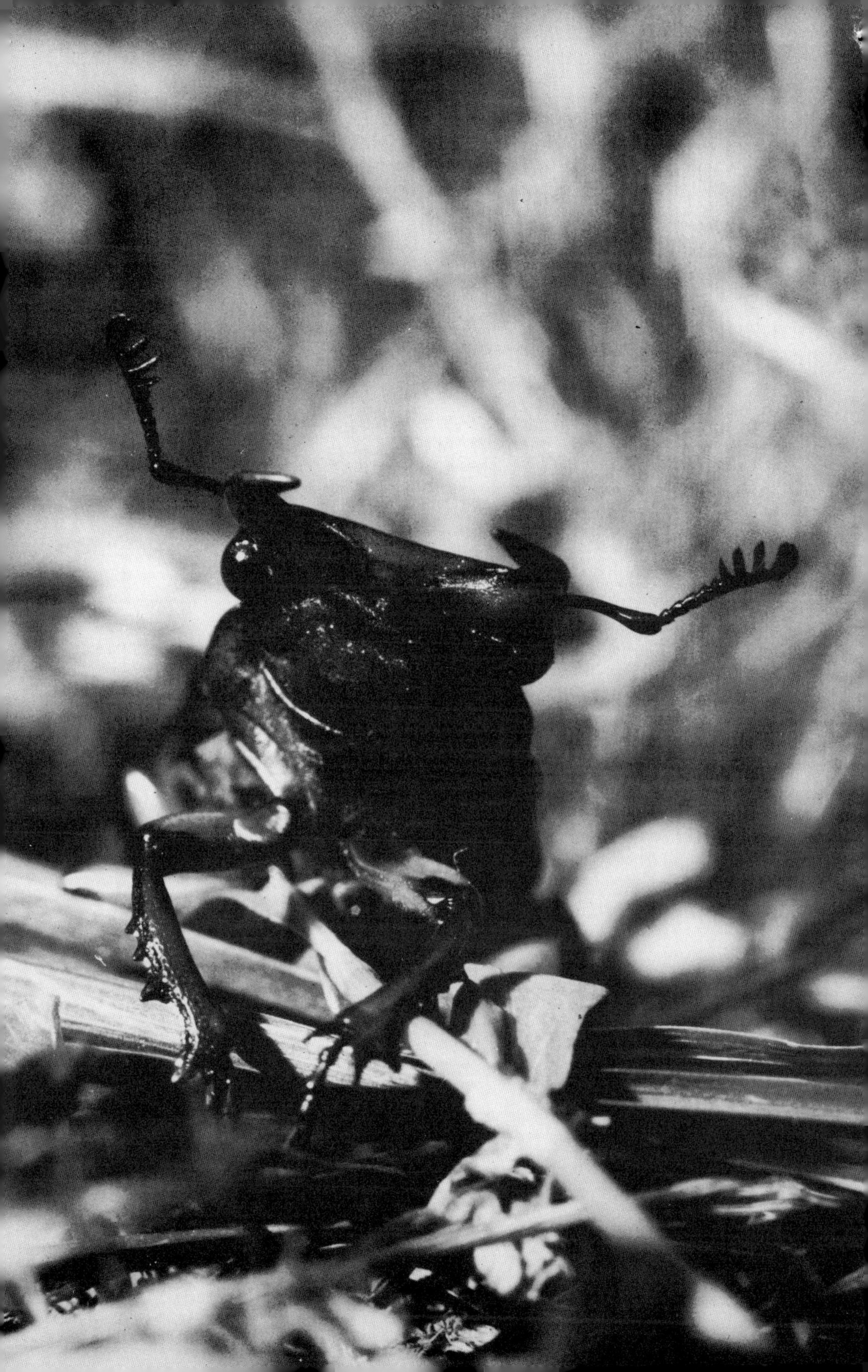

## Moloch Lizard

There are a lot of animals that like to eat lizards. Birds, mammals, snakes and other lizards all prey on these harmless creatures. Lizards, then, have to be fast and they have to be deceptive. This moloch lizard looks and feels unpleasant to the touch (and probably to the tongue) and is doubly aided by its appearance because of cryptic markings and coloration. If you take a functional view of nature, this lizard is truly beautiful.

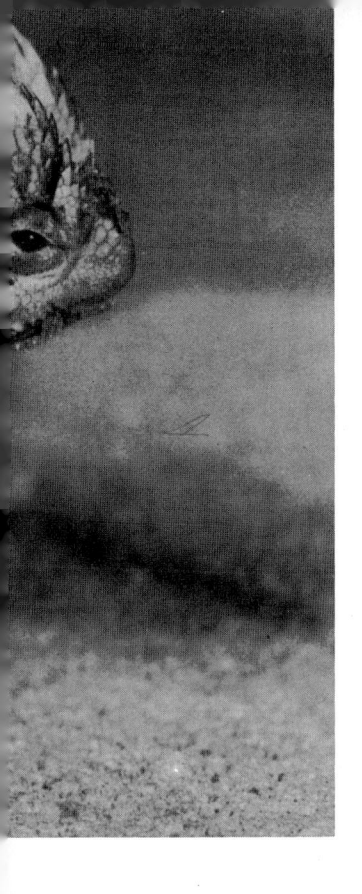

## Ranzania Beetle

Beetles differ from the other orders of flying insects in the design of their first pair of wings. They are changed into shells and cover the second pair, the pair left to them for flying. They aren't as adept in flight as other insects but that shell offers effective protection. We don't even know how many species exist but it is in the hundreds of thousands. The beetle idea has been a good one. They are powerful biting insects, well-protected and diverse as an order. Unfortunately, some of them are among the most harmful agricultural pests on earth.

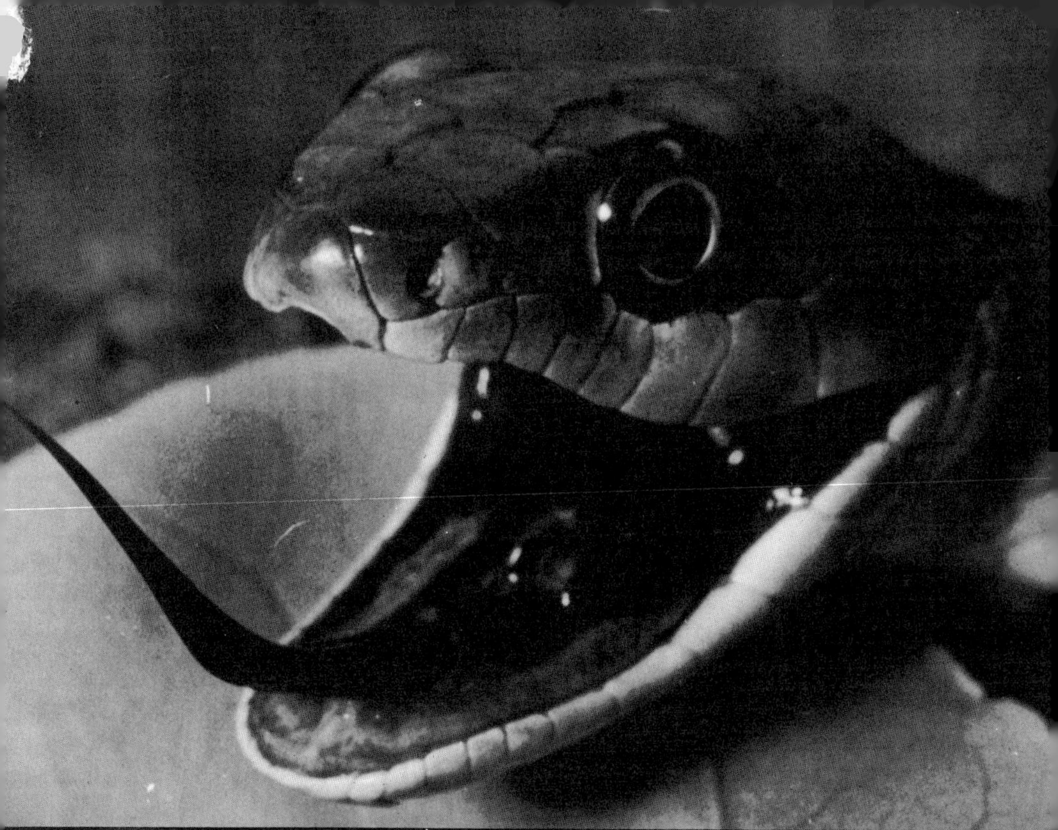

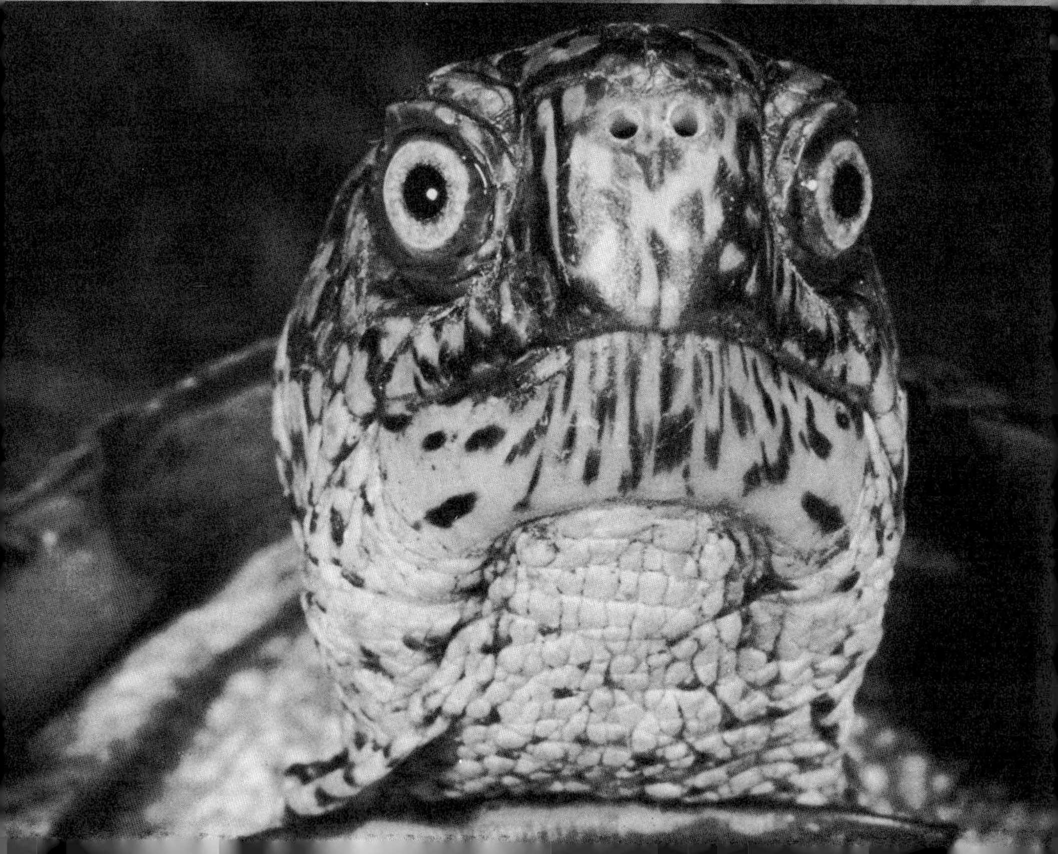

## Green Tree Snake

The reptilian face is one that bothers us. First, there is the fixed stare. As animals that communicate with our eyes and with an enormous number of facial expressions, we are disturbed by animals that have few communications, if any. A turtle doesn't smile, a snake doesn't furrow its brow to register concern, and its eyes don't blink. The fact that the snake tastes the air with its tongue is also something that people find hard to take. The slightly elevated position of the turtle's head when it is looking at something seems to make it appear supercilious. We are, in fact, in trouble when it comes to reptiles. They don't fit our concepts of the way animals should look and behave. In truth it is our attitude and not their form that is bizarre. We have the intelligence to know better than to think that way.

## Box Turtle

#  Chameleon

# Marine Iguana

It took a good many scores of millions of years for natural selection and environmental pressures to give us the lizards we have today. The chameleon's turret eye is a perfect solution to the chameleon's sight problems—as a hunter and as potential prey. The scaled skin of the chameleon and the iguana offer appropriate protection and enable the animal to survive weather, parasites and at least some predators. The spiky crest on the iguana probably intimidates some enemies and probably also serves as a display mechanism during courtship. Feature by feature, these animals can be seen as the culmination of millions of probing ideas over millions of years in an infinite number of situations. We will never know about the bad ideas that failed.

## Grasshopper

Our prejudice plays cruel tricks on us. There are billions and billions of grasshoppers born each year and they are generally good to eat. They have been eaten, in fact, by people as far back in history as grasshoppers go. They are still eaten as delicacies in parts of Asia today. Yet, even with food shortages, we would not sit down to a platter of boiled or broiled grasshoppers. We only eat beautiful things like crabs and lobsters and shrimp. Tastes and attitudes are strange phenomena and we sometimes pay an awful price for having them.

**Praying Mantis**

In an old book, I've forgotten how old it is, I read a traveller's description of the animals he encountered in some exotic country. In one place he really got carried away. He had used every adjective he knew and wanted to place the ultimate onus on an animal that displeased him in some way. He said it was inhuman. I guess that is what it really amounts to, much of the time, we find animals inhuman; that is to say they are unhuman—and that is supposed to be bad. What an absurd concept that is and what a poverty-stricken view of the miracle of life on earth.

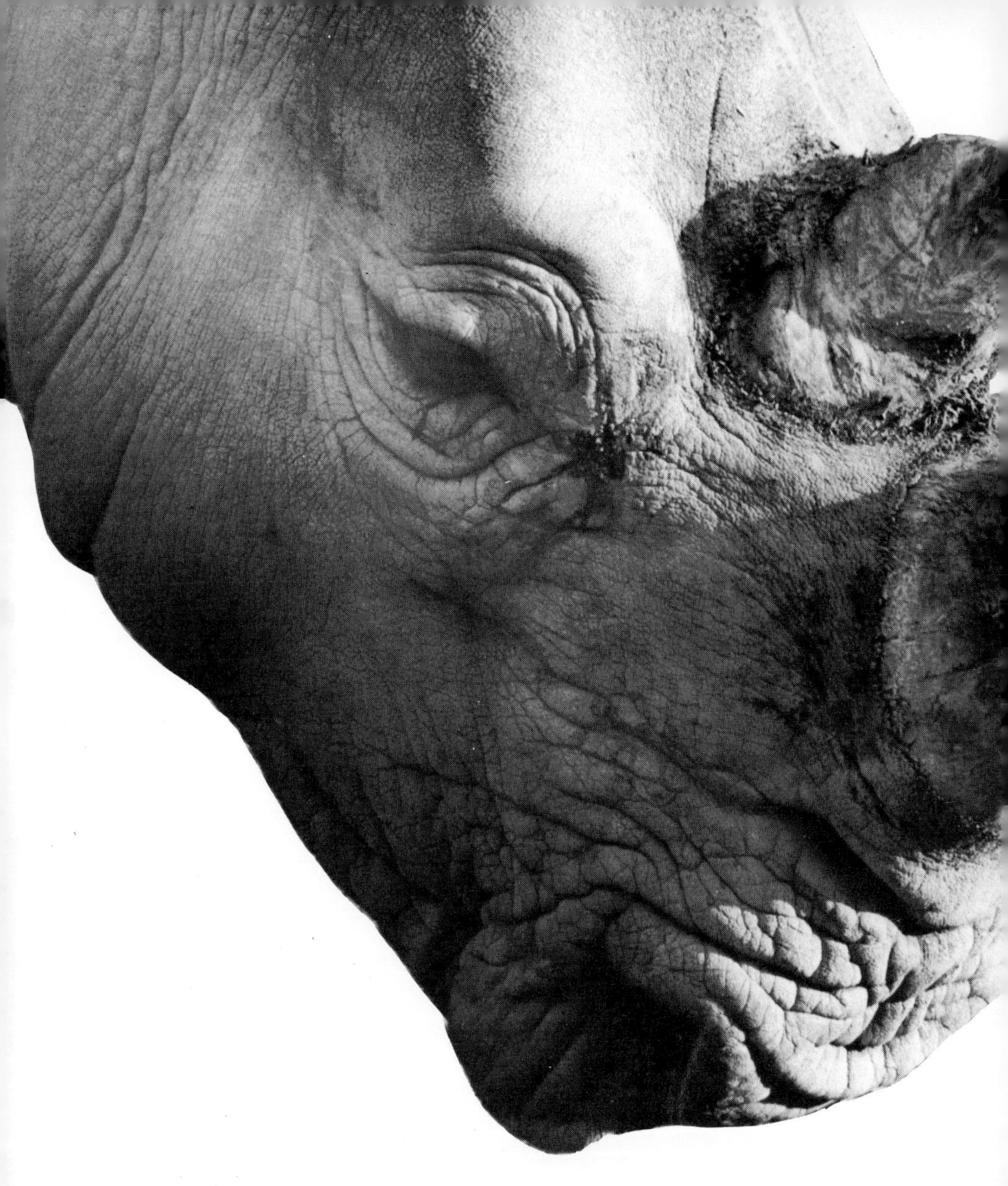

## Rhinoceros

Some animals look absolutely prehistoric. It isn't a rating or a value judgement of any kind, but rather the feeling an animal gives of coming from a different kind of world. That is basically true in the case of the rhinoceros. It is a relic, a holdover of the great Pleistocene fauna of a million years ago and more. That makes the rhinoceros all the more precious, all the more important to preserve. As it appears, it is an exceedingly ancient animal, a reminder of what life was like before man was man and killing was fun.

The more nearly like us an animal is the more we project, of course, and the more interested we become in their sometimes bizarre appearance. A beetle can look like anything it wants to and all we take is passing notice, but a baboon is something else again. It is almost "one of us" and therefore its fur becomes a *hairdo* and that is followed immediately by all manner of anthropomorphic projections. I have seen more than one person stand in front of a monkey or ape cage and follow the animal's yawn with their own. The primates involve us. We are that close.

**Yawning Hamadryus**

Not only do we make "little people" out of our fellow primates we then try to interpret their facial expressions in human terms. It is an all-out attack on their monkeyhood. Take this male stump-tailed macaque, what is he "saying?" Is he laughing at us? Is he expressing pleasure or displeasure at something the photographer has done? It is impossible to say, but he may have been evaluating the smell of a female in heat. His expression may be one of terror or territorial claim for his troop. In fact, despite our insistent opposition to the idea, he may be doing something very monkey-like and not human at all.

## Stump-tailed Male Macaque

**Proboscis Monkey**

We see some animals as not only miniatures of ourselves but as cartoons. The large and not very gracefully shaped proboscis monkey from Borneo with the large pendular nose (only males have them) indeed appear to be misshapen men. These interesting animals are becoming increasingly rare and it isn't because of predation by enemies like the clouded leopard. Man is taking away from them their traditional riverbank forest home and there is unwarranted direct destruction as well. If we keep on as we are, the sad faraway look we might project on this picture of an adult male specimen might take on new and far truer meanings.

## Young Screech Owls

Owls are nocturnal creatures and are designed to hunt when light is at its lowest point during the twenty-four hour cycle. They are also designed to kill quickly and efficiently. As such specialized creatures, owls don't always fit our concept of the beautifully feathered creature. Baby owls are even stranger looking, but their down is the perfect protection against the elements during their brooding weeks before they attain juvenile flight and later adult plumage. The large eyes of these young birds in their already discernible facial discs give some hint of the kind of life they will one day lead. The evaluation of their temporary masquerade-ball appearance is, once again us, not them, at work.

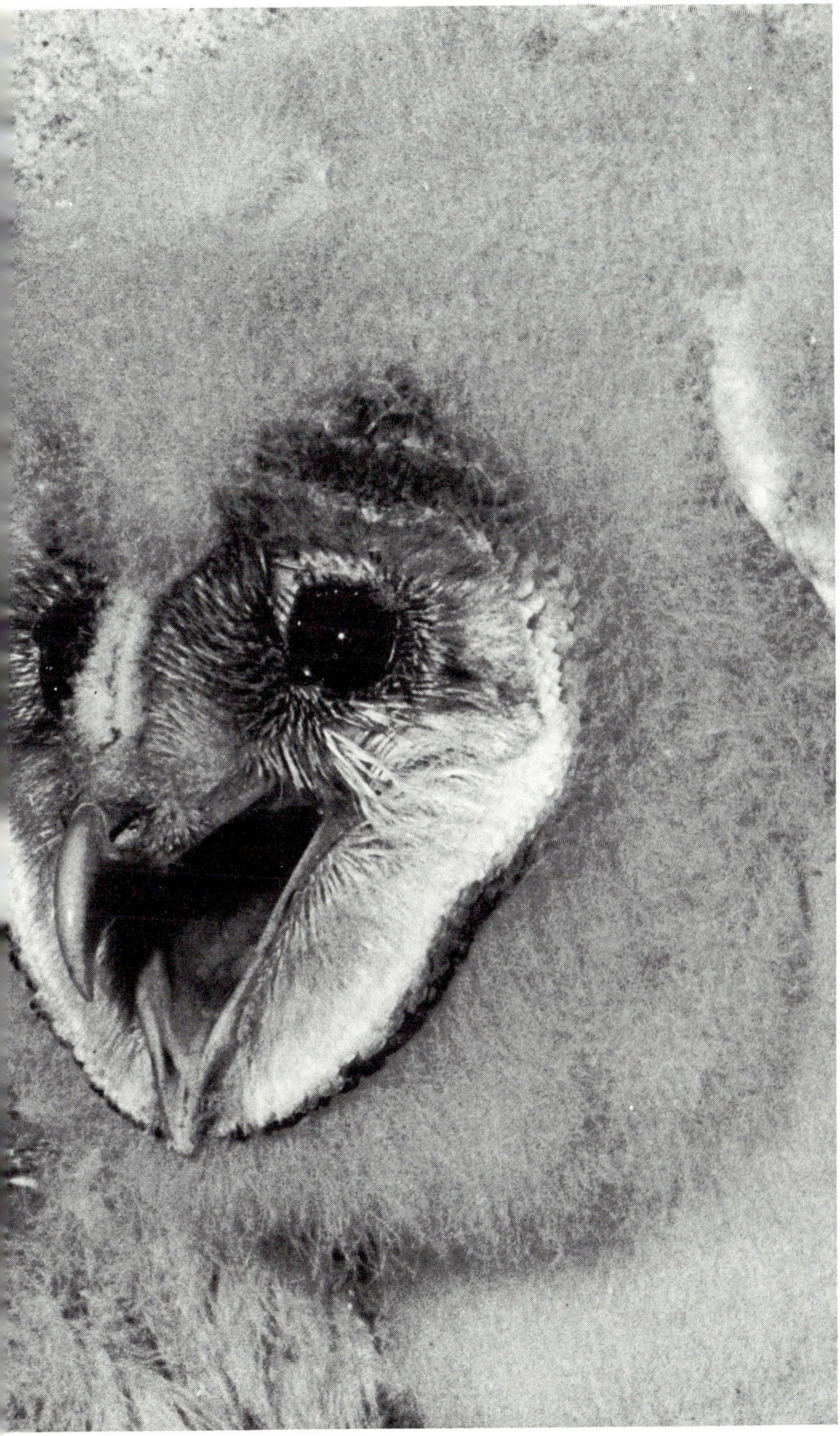

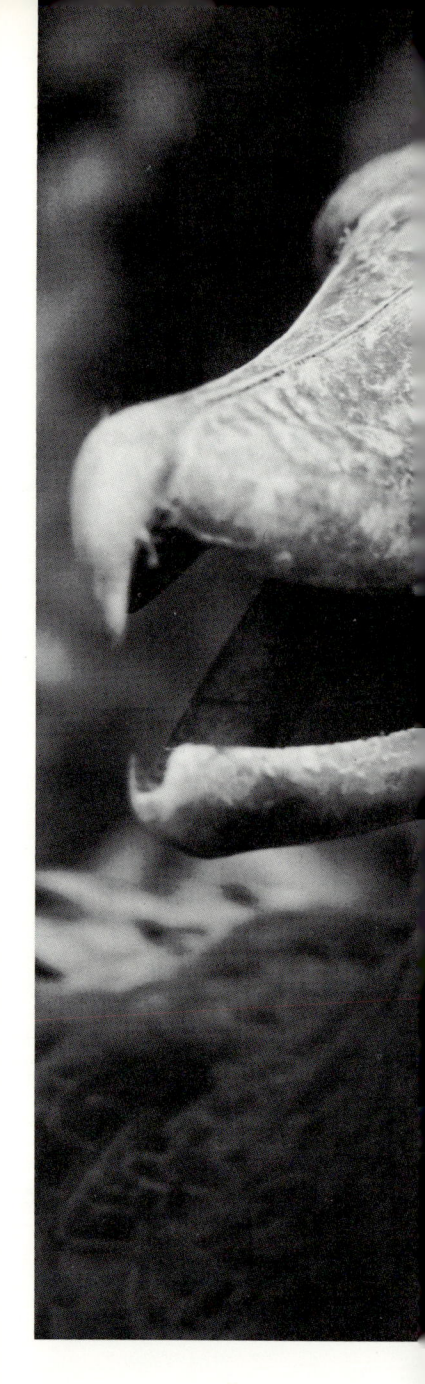

**Major Mitchell Cockatoo**

**Ostrich**

**Shoebill Stork**

Avian facial styles: Upper left, from Australia, the striking Major Mitchell cockatoo and below left, the ostrich. The larger right hand picture is of a shoebill stork. Which of these is beautiful, which homely, which absurd and why? How could we make such judgements and on what basis? Each of these birds evolved in nature in the form best able to cope with the problems of its environment. The large eyes of the ostrich were needed by a bird that could not fly yet lived on open grasslands also inhabited by lions, leopards, hyena and hunting dogs. The cockatoo needed that hooked, powerful beak to aid in handling the vegetable matter on which it thrives. In contrast is the enormous bill of the shoebill or whaleheaded stork. It is ideal for life in an African papyrus swamp. How is one of these adaptations more beautiful than another? Isn't the ability to survive and reproduce the most beautiful thing of all?

29

## King Vulture

The vultures, as ordained by nature, are carrion eaters and that means they must deal with what is to us an unsavory diet. Yet, vultures are very clean birds. Nature has helped them to this end. Vultures have featherless heads and necks. It is easier for them to keep clean when they thrust their head into a rotting carcass because they are not fully feathered. Vultures bathe often and sun themselves regularly. Here again, is a worthy adaptation to a special way of life. It is a miraculous theme throughout all of nature.

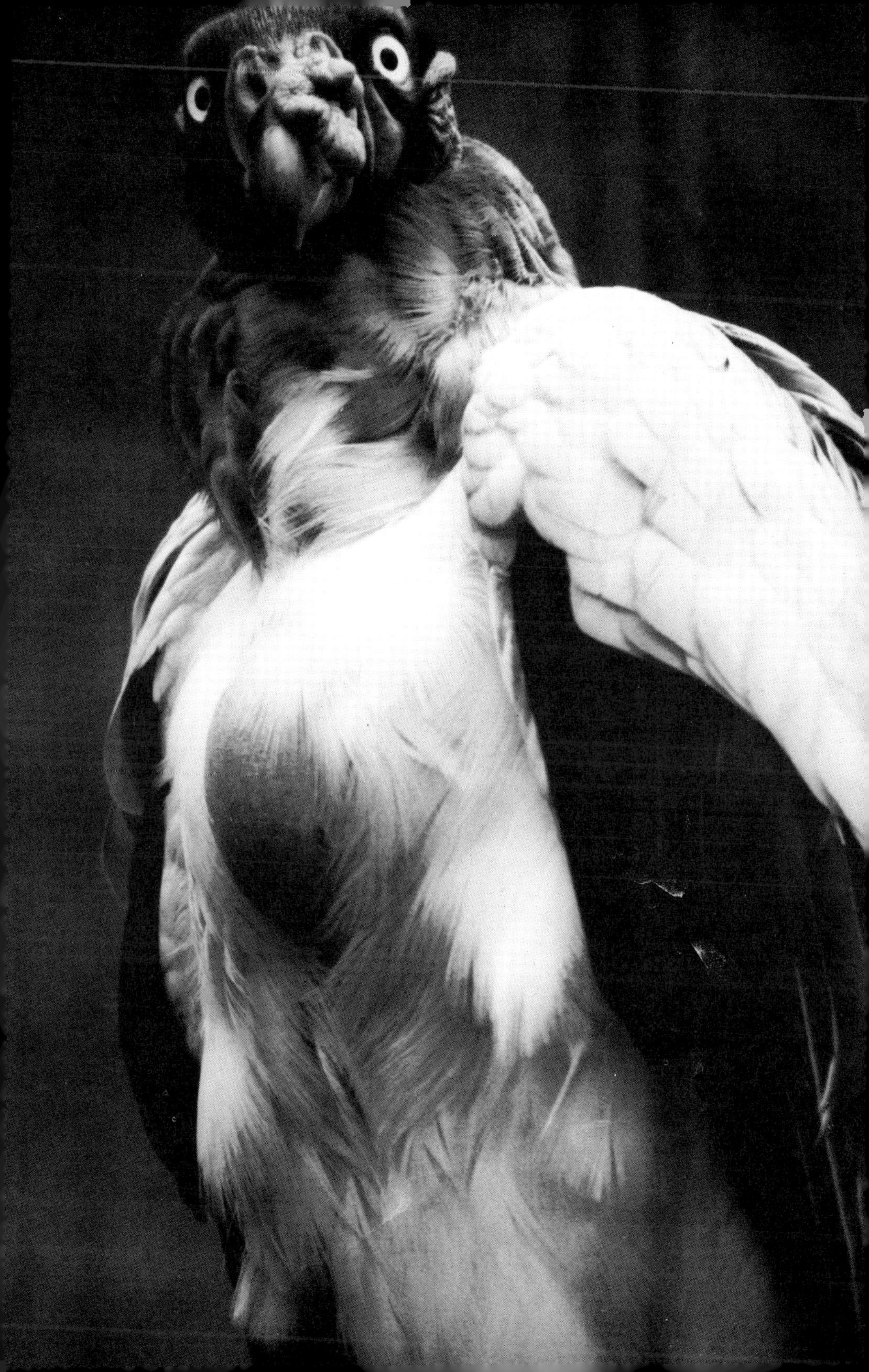

**Australian Pelican**

### Saddle-Bill Stork

The point-of-view of the photographer has a great deal to do, of course, with how an animal will appear to the uninitiated. These two birds live in, near, or on the water. The Australian pelican dips when it is on the water, or dives when it is airborne, to scoop up fish in its bag-equipped bill of impressive size and style. The saddle-bill stork, an African species, is a wader and has a bill that is also impressive. The head-on views selected by the photographers have given these large and important birds a humorous aspect. That is, at least in part, because we equate the bird's bill with the human nose. It is much more than a nose though, it is a mouth, a hunting and nest-building tool and very often an important device in courtship and combat. A camera angle can't take away from all that!

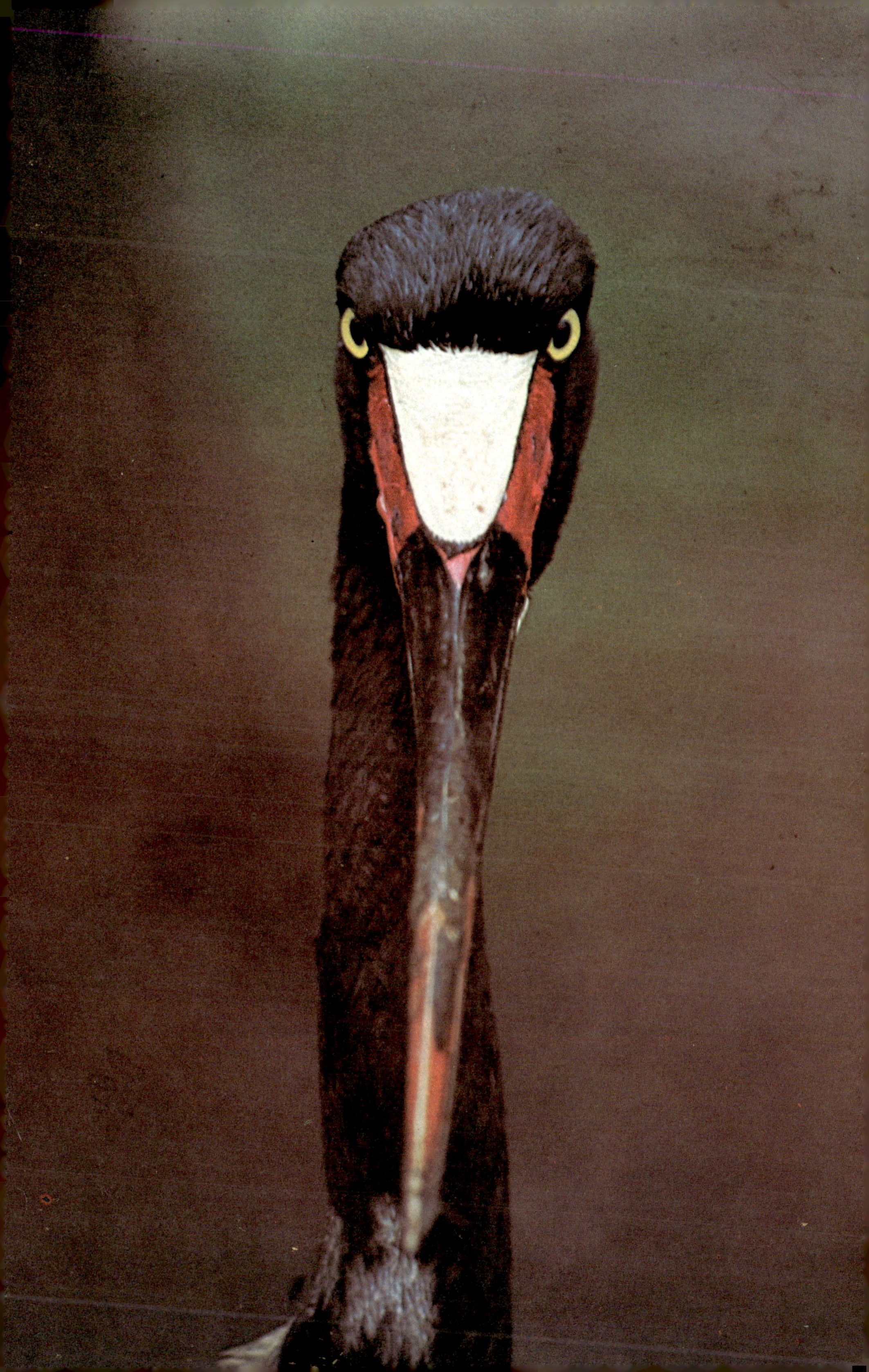

**Rockhopper Penguin**

We develop fixed ideas of what animals are supposed to look like and even how they are supposed to behave. When animals refuse to sit still for our projected images, we describe them in slightly or even openly derisive terms. One group of animals that has deviated yet survived is the penguins. We think they are "cute"—and to a purest that could be as damning a word as bizarre, weird, savage or any other inappropriate appellation. Penguins don't fly, they are better adapted to the sea than to land and their wings have been converted into flippers that seem to be hairy rather than feathered. They are unbird-like birds, yet we have forgiven them. *Cute,* apparently, allows for a lot of deviation.

## Ant Curtain

When ants span open areas by forming bridgework or a curtain with their own bodies, we see a form of behavior that troubles us. The selfless mass behavior of the social insects (bees, wasps, ants and termites) is something that seems nightmarish to us. The loss of individual identity is horrifying. It makes our "skin crawl." There is even a medical name for that sensation—formication.

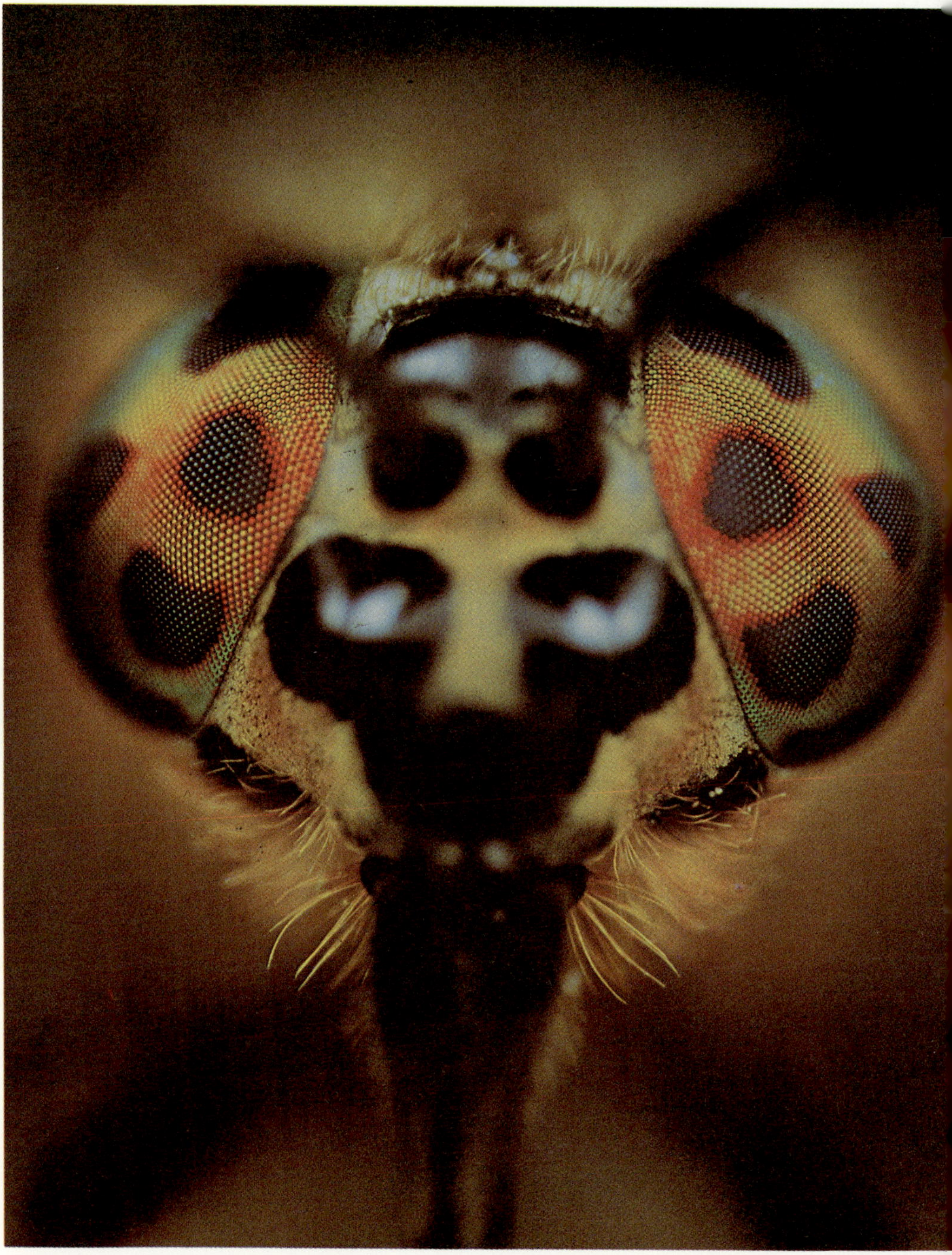

## Head of Fly

Down, far below our normal visual range, there exist details of animal structure that are as fascinating as they are remote. Here are three close-ups that show some of the world lost to us when we don't have a magnifying glass at hand. Far left is the head of a fly with the marvelous multifaceted eyes that can never see as much beauty as they themselves possess. At left is a close up of the eye clusters of a jumping spider. What must it be like to look out through eight eyes or, as in the case of a fly, a single eye with hundreds of lenses? At right is a profile of a member of the Euchromiidae, day-flying moths that some people confuse with bees and wasps.

## Jumping Spider          Euchromiidae

What would it be like to have vision that could focus down to such sights as these unaided? How unfamiliar even our own backyard would become. We would be a stranger and the strangest things of all would be the things we have lived with all our lives. Focal length, and our ability to see are what make animals seem bizarre, not just their structures or habits.

**Mandrillus Sphinx**
Again a primate and one of the most colorful of all, the splendid mandrill is a native of Africa. He is one of the largest and most aggressive of all the dog-faced monkeys. The males display their marvelous facial patches and we can only guess at all the things they mean. They are signs of a mature male when they are in full bloom; they are probably attractive to the female at mating time. They probably play a role in aggression or countering aggression, for they make the big male seem even more formidable. They have evolved as a structure, to be sure, but as a structure that is an important part of a way of life. It is all coordinated, all part of a successful form of life and none of it, structure or behavior, can reasonably be considered apart from the rest.

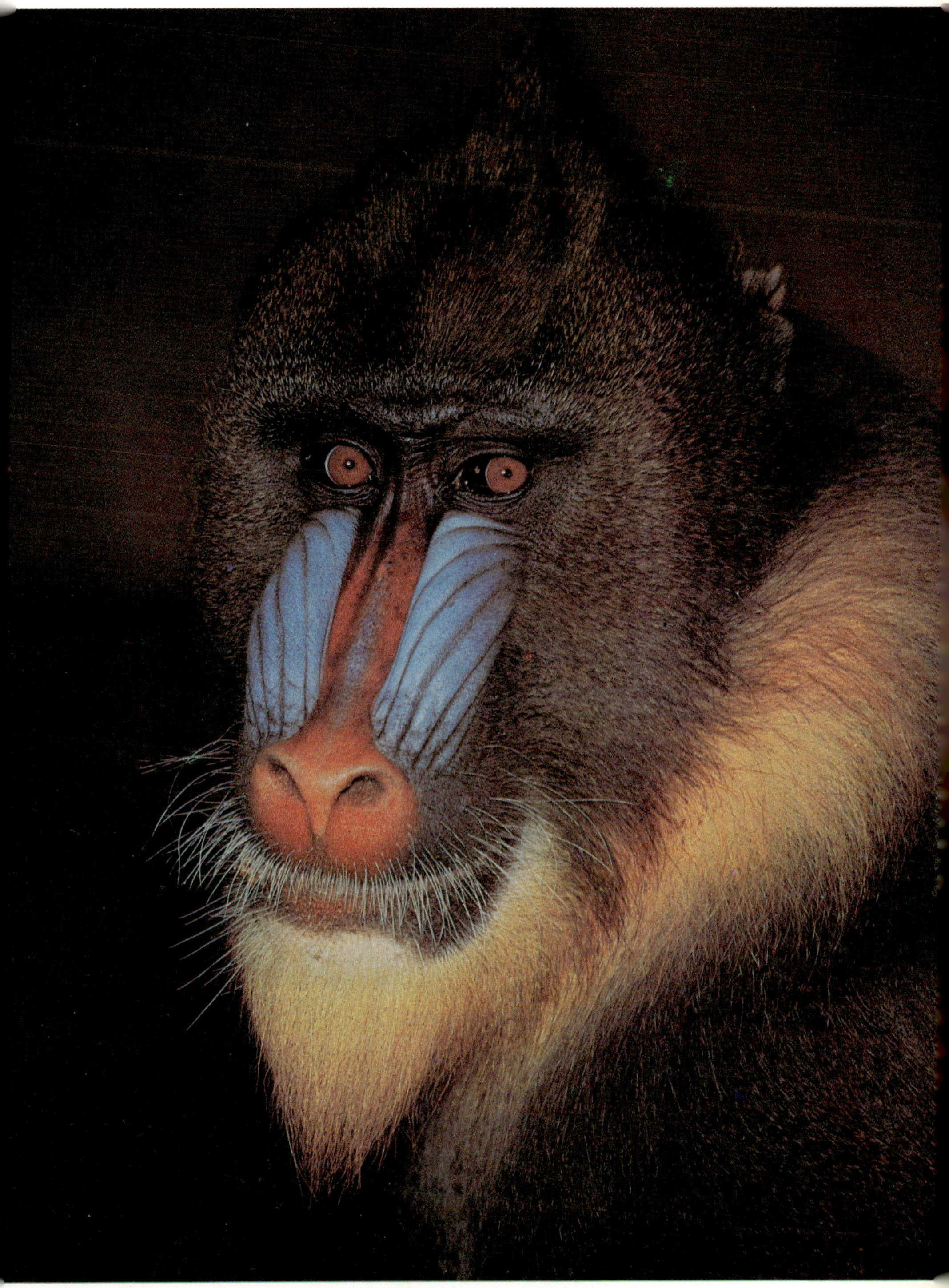

Frogs, since they spend much of their time largely submerged, must have eyes that can protrude above the water. There are raccoons, weasels, birds of all kinds and other predators as well that would have a frog very quickly if it could not see them coming and submerge and vanish into the bottom debris. The extended or stalk-eyed look of the frog would be most unattractive in, let us say, a zebra or leopard, but in the frog it seems right. It is right, of course, not because of how it looks to us but because it spells survival for the animals themselves. The extended eyes of this Asiatic horned frog have seen a lot of cranes and herons in time for the animal to escape their deadly beaks.

## Asiatic Horned Frog

Eyes again: another advantage of eyes that protrude from the skull—peripheral vision. Both the chameleon and the frog pictured below—a reptile and an amphibian respectively—can rotate their eyes through a greater field than we can. They are so vulnerable to attack that they need these broadened visions to keep track of approaching hazards. Since they both seek insects the roving eye also helps locate prey alighting nearby where the tongue can retrieve them.

**Chameleon**

**Frog**

# Deep-Sea Angler

In the sea, living creatures take on shapes even less familiar than the strangest animals of the land. Their world of cold, pressure and dark is foreign to us and so are the forms that can tolerate these nocturnal influences. Creatures of the sea are not equipped for change as are their counterparts on land. Temperatures are nearly constant (they *are* constant at great depths); there is no night or day and no change in radiation from the sun or space, and there is never flood or drought. This creature from the deeper places in the sea we find bizarre enough to be grotesque. It is a voracious hunter of what we assume is a very ancient design. The deep-sea angler moves through a world in which man cannot survive, nor can he really understand.

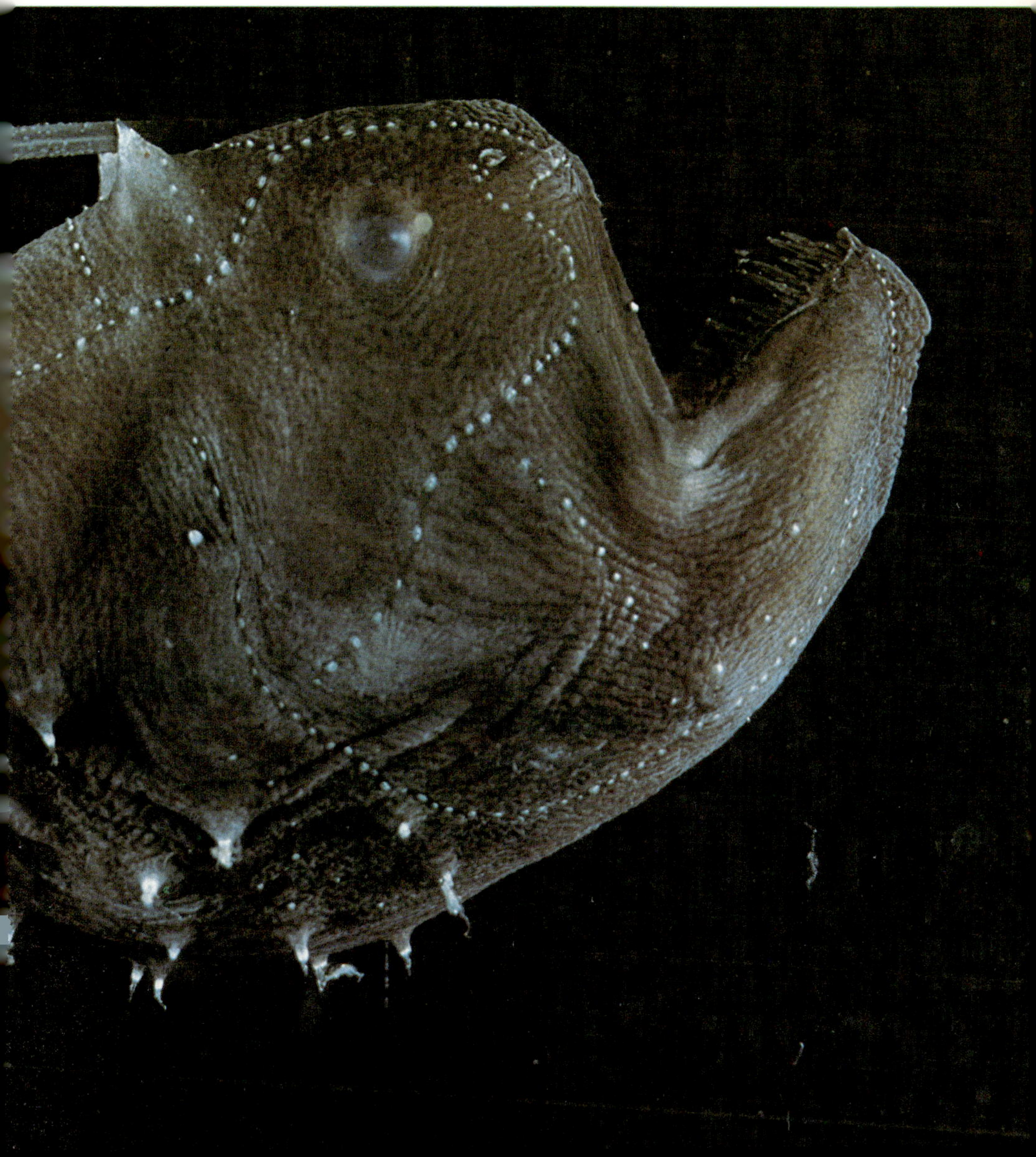

## Millipede

The centipede (the creature of "a hundred legs") has one pair of legs per body segment. The millipede, on the other hand, pictured here in a coiled defensive position (a creature of "a thousand legs"), has two pairs of legs per body segment. There is another important difference, too. Some centipedes, at least, are venomous and some few may be dangerously so. All of the glistening multisegmented, many-legged millipedes are harmless.

45

### Deep-Sea Squid

The squid is a mollusk without an external shell. It is a creature of the oceans, of course, and the species pictured here is from the deeper regions of the sea. The squid and its relative the octopus are probably the most intelligent of all invertebrates and they are certainly the largest. They are swift, sure hunters and are designed for a way of life we can barely imagine. This squid travels where even our nuclear submarines cannot go without being crushed. They tolerate temperatures every moment of their life that would still our hearts in minutes. Their vast eyes gather light so thin it would be invisible to us and their many arms and two tentacles are equipped to reach, to grasp and to hold. We may not think the squid aesthetically pleasing but we must admire the perfection of its design and function.

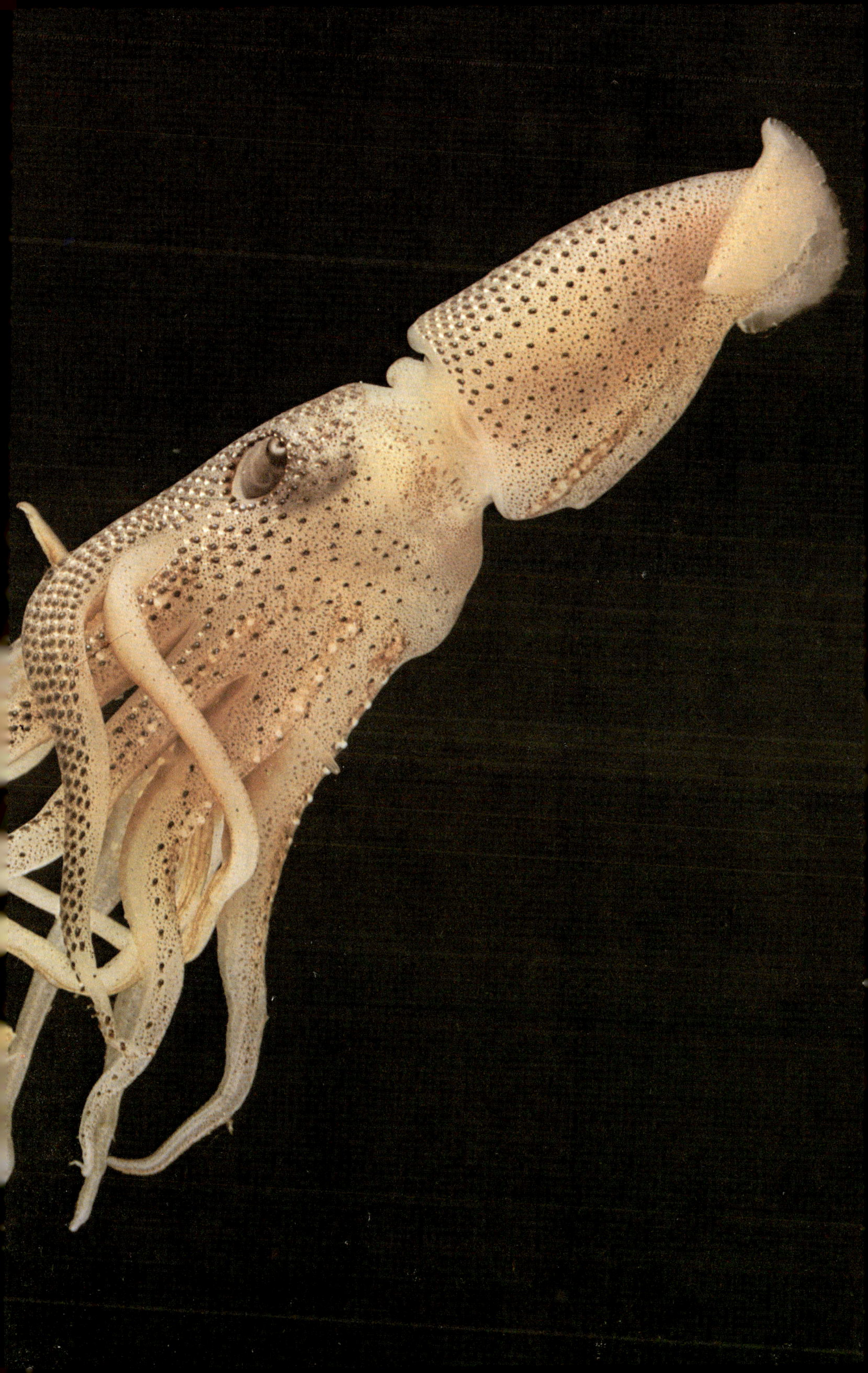

## Yellowfin Toadfish

One way to keep from being eaten is to be of such a shape that swallowing you becomes at least difficult and perhaps impossible. To the fish-eating fish the toadfish at left and the newly inflated puffer at right are not a convenient size or shape for swallowing. As for the markings, since we cannot see with a fish's eye in the fish's world with the fish's intent, we don't really understand them at all. What seems brilliant to us may be cryptic to fish and what seems strange and unaccountable in our view may be madly attractive to fish of the same species but different sex. There may be other factors as well that we can hardly guess.

**Puffer Fish**

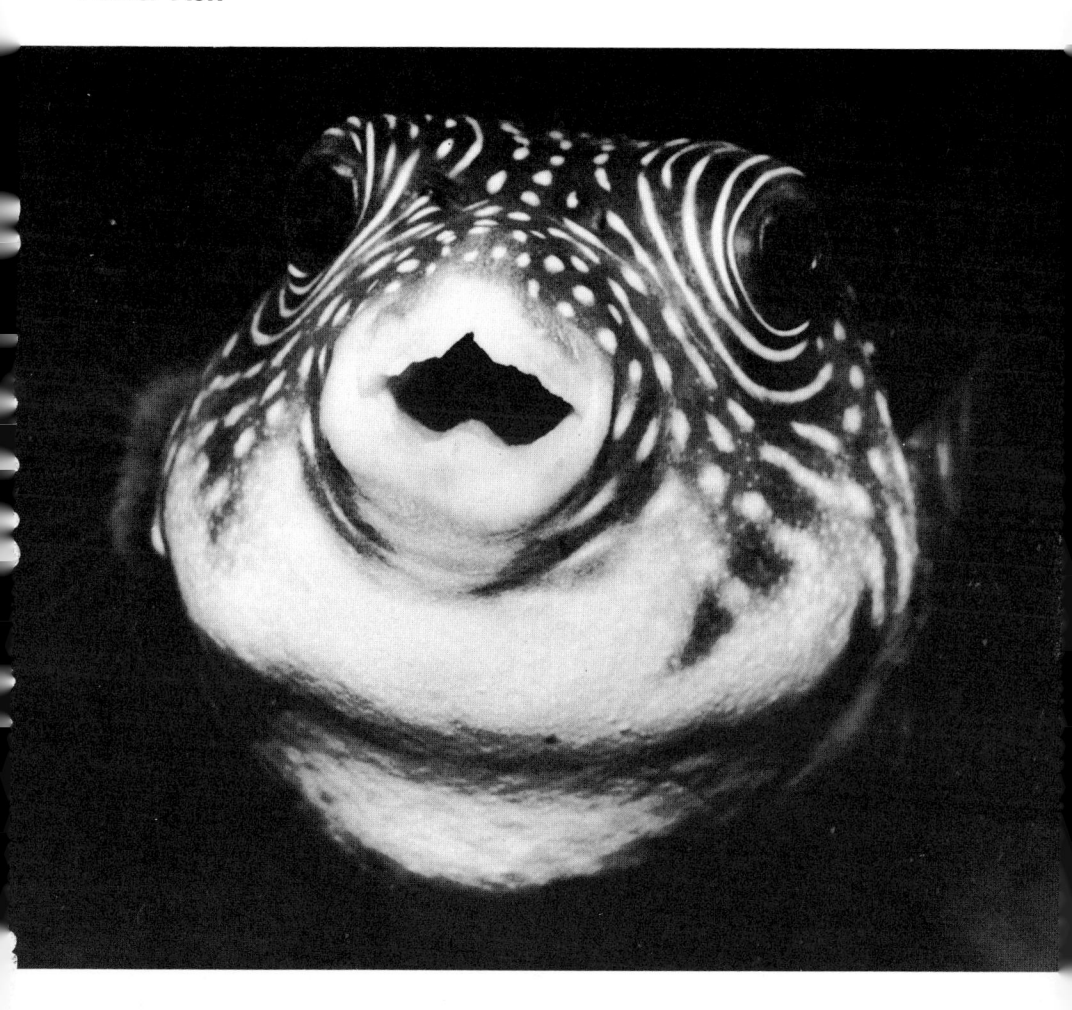

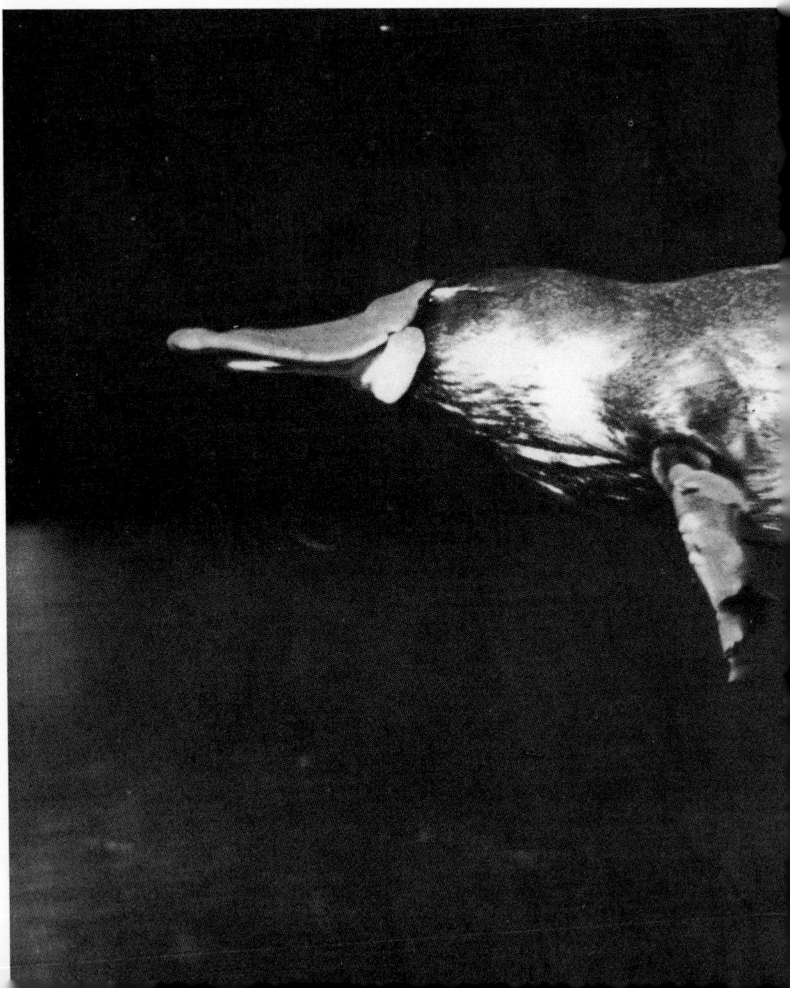

# Platypus

The platypus is a mammal. The platypus is a mammal? That can be phrased as a statement or almost as a question. It is a mammal, but it is as divergent from the mainstream of mammalian development as you can get. It is often called the duck-billed platypus and some people take the *duck-billed* part of it to suggest an affinity with birds. That isn't accurate, for the strange frontal appendage is only superficially duck-like and suggests no affinity. This mammal does lay eggs, however, and that is about as different as a mammal can get. The female supplies milk to the young but there are no nipples as such. One other thing, the platypus has a venomous *kick,* at least the male does. There are venom glands and venomous spurs on the rear ankles. We don't even know what they are for. This is without doubt the strangest mammal of them all.

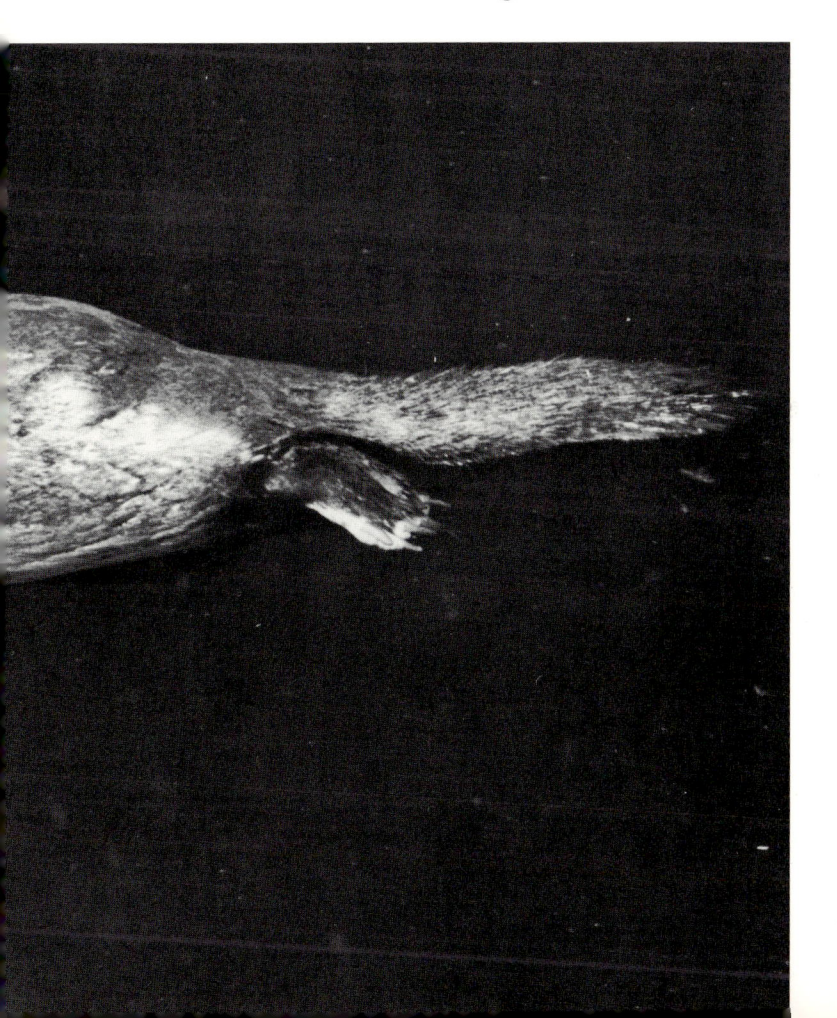

### Peacock Worm

Some words carry feelings as well as meanings—we say they are connotative as well as denotative. Surely, the word *worm* carries feelings as well as defined meanings. Worms eat us when we are in our grave, worms crawl in dark places and must surely be slimy and ugly. But, what happens to us when we come upon a peacock worm such as that seen here? What happens to our mental set? This feather-like finery can belong to a worm as surely as any other creature and we find ourselves examining our own attitudes and feelings and the things we think we know but perhaps do not.

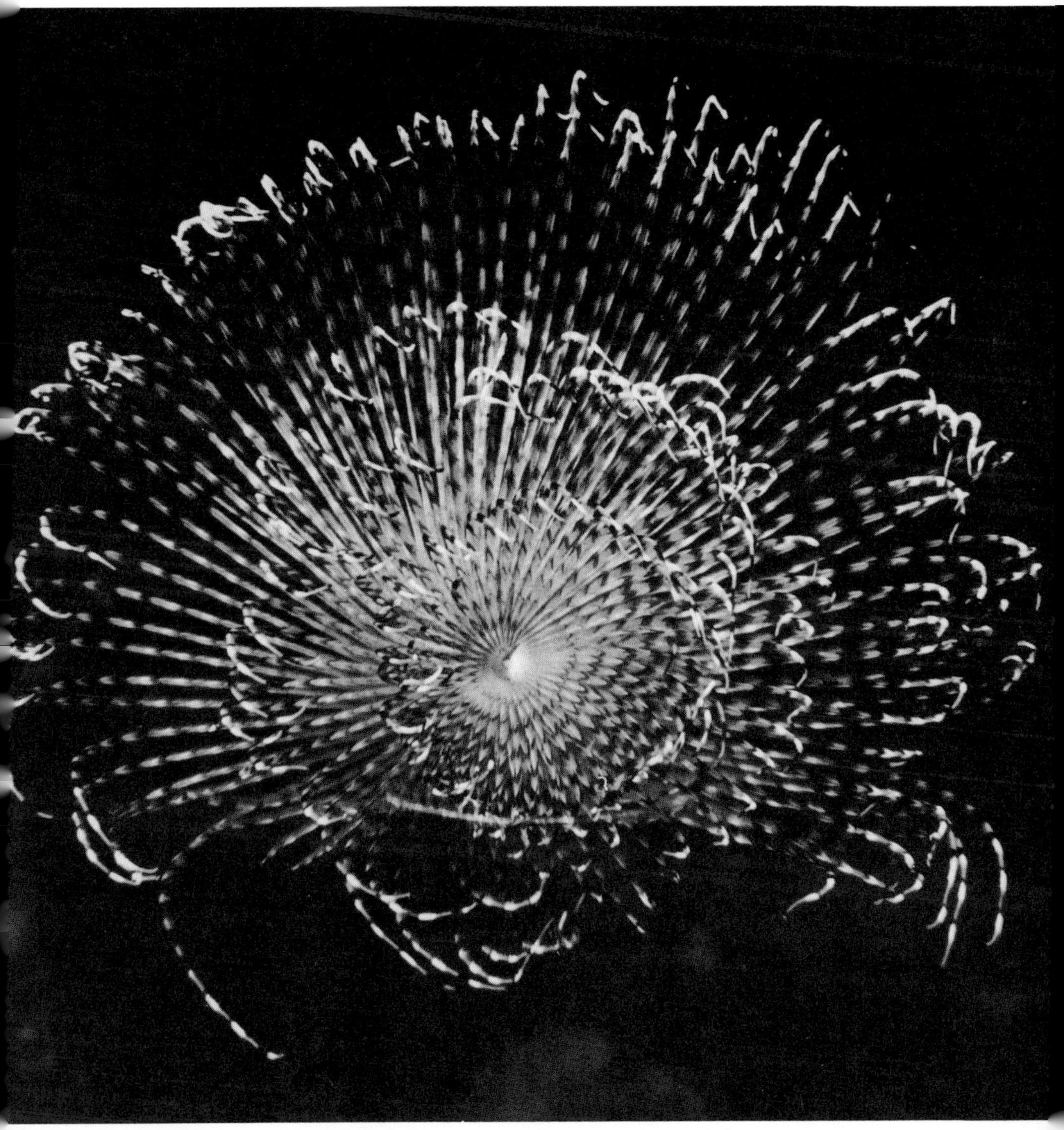

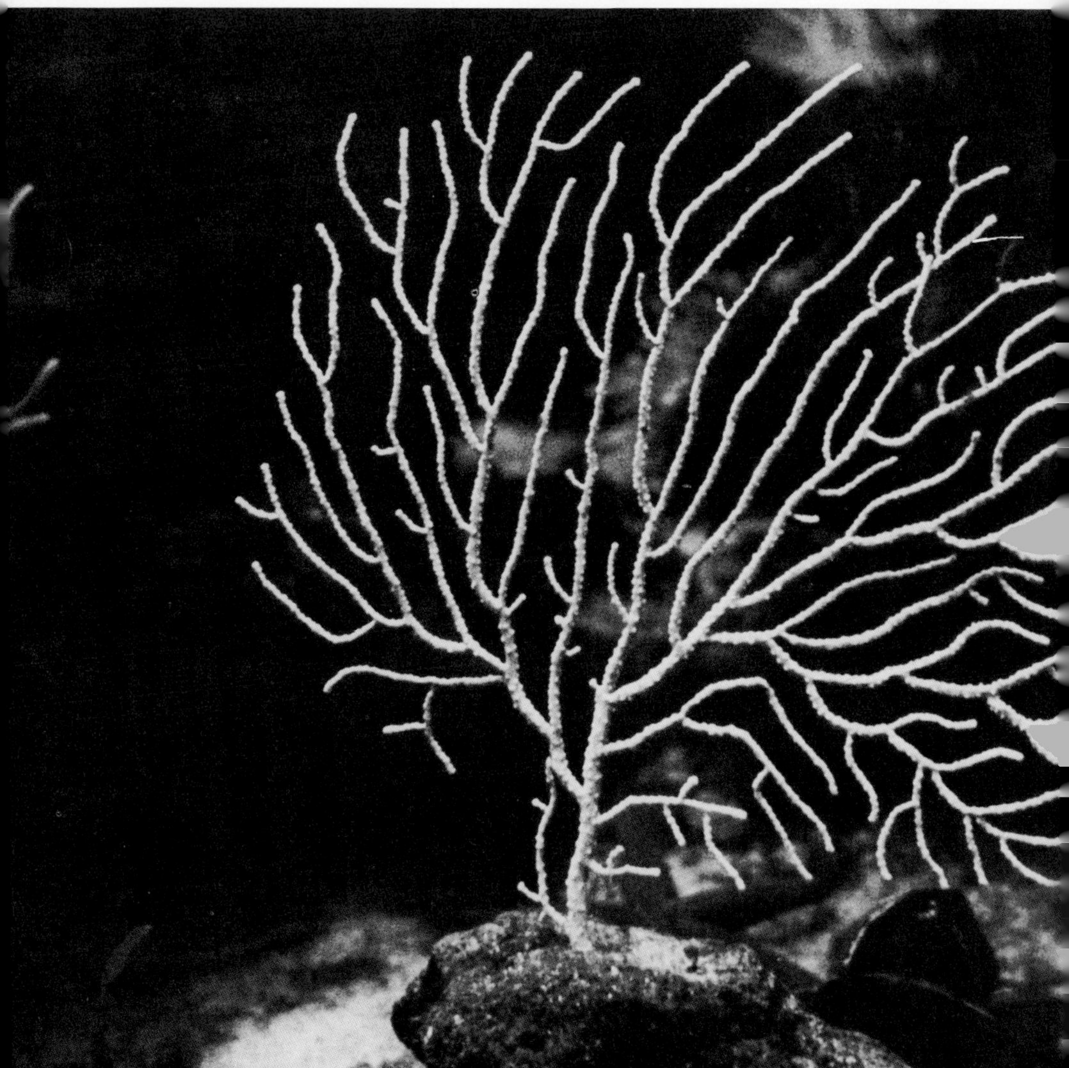

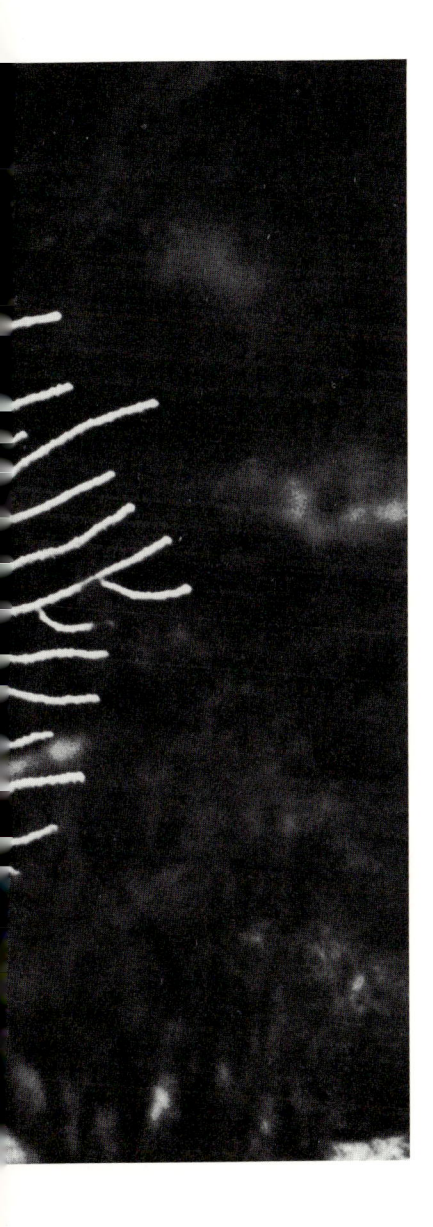

## Violet Sea Fan

Some animals double-cross our mental conditioning and take on shapes we can barely allow. This is not a plant, but a colony of animals. The creatures that created this violet sea fan are related to the corals and the anemones. The colony does not take on the form it does so that it will resemble a plant, for no creature with which it shares the floor of the sea or the side of a reef could make that kind of a comparison. Only we can. Long ago the coelenterate species responsible for this enchanting underwater form evolved it as ideal for tasting the currents and extracting food for each member of the condominium.

## Zebra Moray Eel

The moray eels are shy and retiring creatures who deeply resent intrusion. They are not venomous as is so often claimed, but they do bite and bite hard when disturbed. Eels in general are snake-like to us and therefore suspect. We are very hard on creatures that are elongated and apparently legless. The striking markings on this moray may serve a number of purposes. One of them could be to warn away potential enemies to avoid direct confrontation. A moray eel will bite hard but it will avoid it if possible.

## Porcellanid Crab

Few animals combine features ugly (to us) and beautiful more effectively than the crabs. The crustaceans, of course, are related to the spiders and insects. The crabs are members of the order Decapoda which includes, as well, lobsters, crayfish, shrimp and prawns. Many of these animals are among the choicest of human foods. For some reason we find crabs humorous. We don't laugh at lobsters and shrimp, but crabs assume a character we can only call cartoon. To use what is now a perfectly legitimate English term, crabs are disneyesque. They are also ancient and successful hunters in the sea.

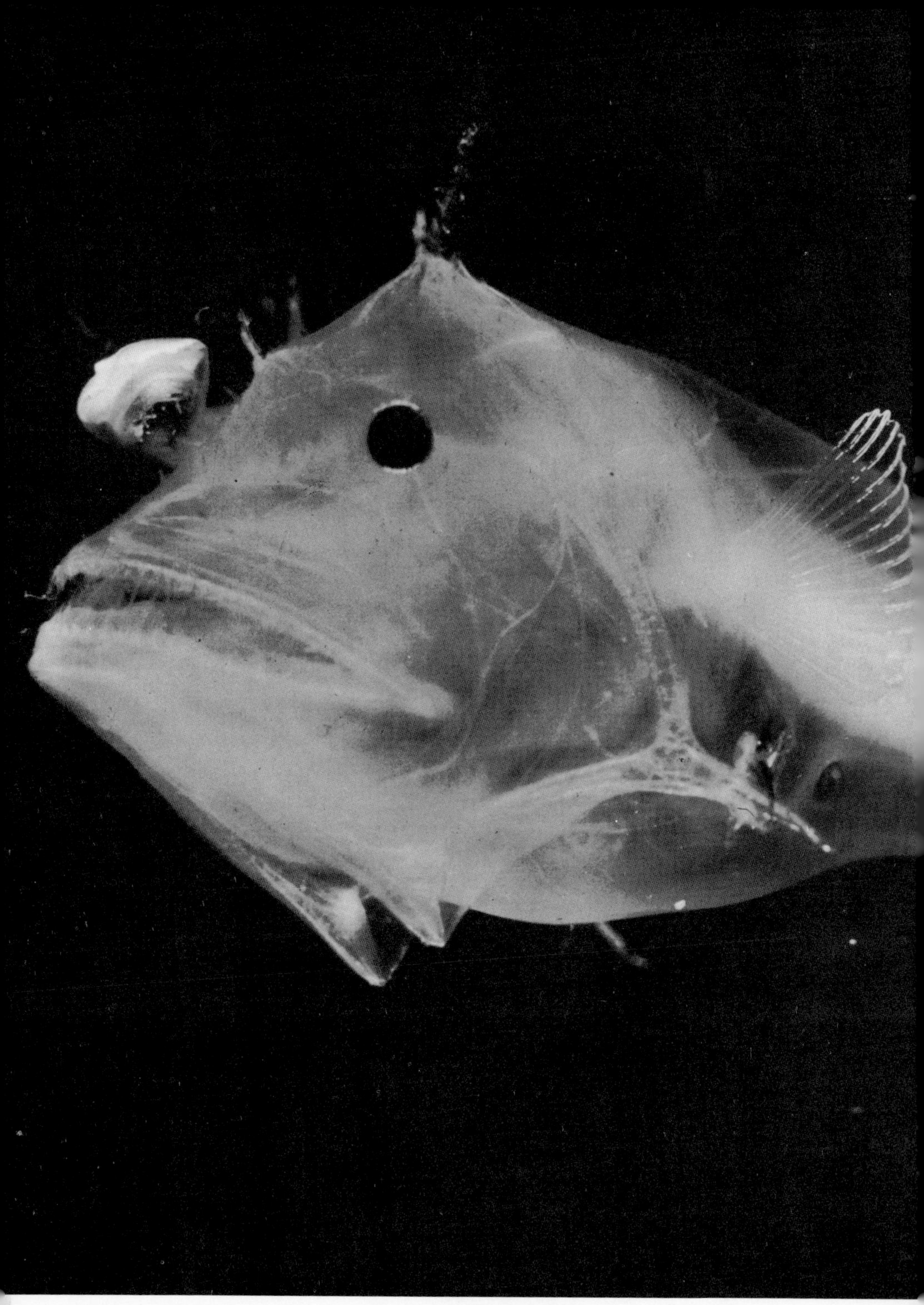

## Deep-Sea Angler

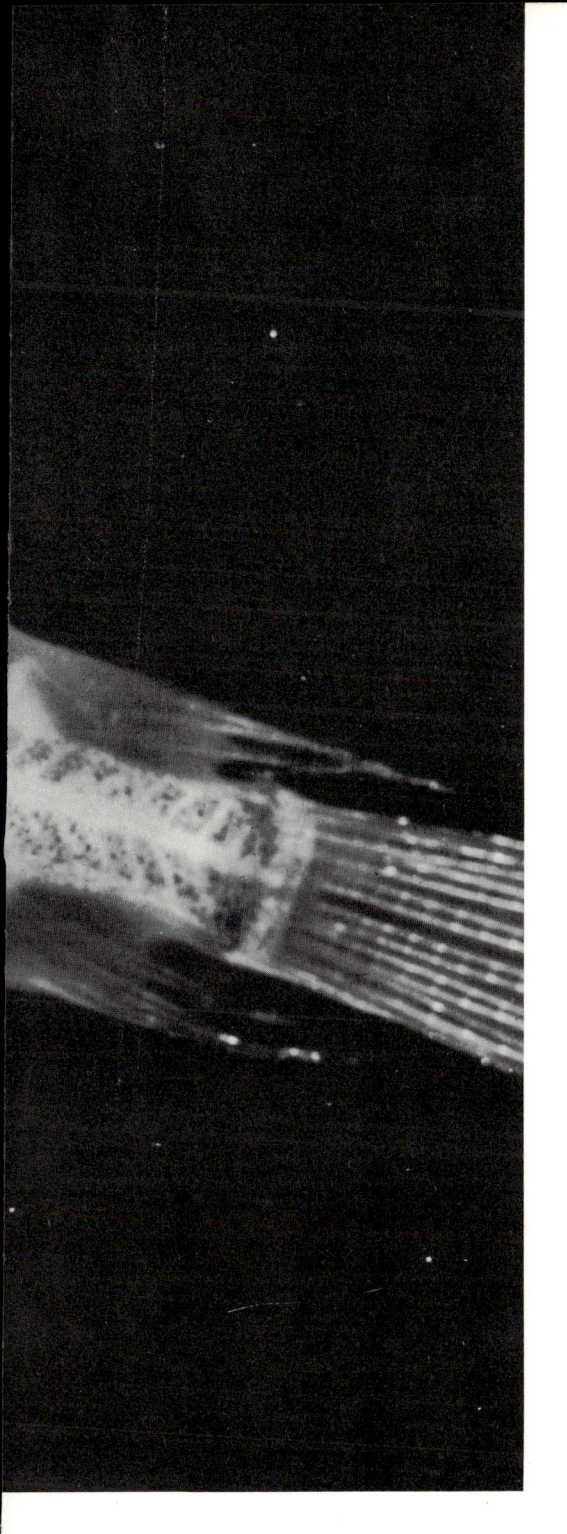

The deep-sea angler in its realm of eternal night is uneffected by the crushing weight of the ambient sea because the pressure inside of its delicate tissues is the same as outside. It is in a state of equilibrium. Its most ancient ancestors undoubtedly evolved nearer to the sun but competition and unexploited opportunities caused some to seek prey and safety for themselves deeper in the sea—ever deeper until the strange form we see here evolved. The journey toward the center of the earth was done at a leisurely pace and so were the physical changes required to accommodate the conditions that the animal had to face.

## Wolffish

There are, of course, no absolute values contained in the concept of beauty. (The Greeks developed canons of beauty but they involved only the human face and body, but had nothing at all to do with fish.) There may be, however, absolute values in the concept of ugliness—subjective though they must always be. Who would say that the wolffish pictured here is not ugly? You may think that it is a nice ugly, but ugly it is. That wasn't the design's impetus, but that is the outcome as witnessed by us. What was the impetus? The same as for all animals—to be better at eating than at being eaten.

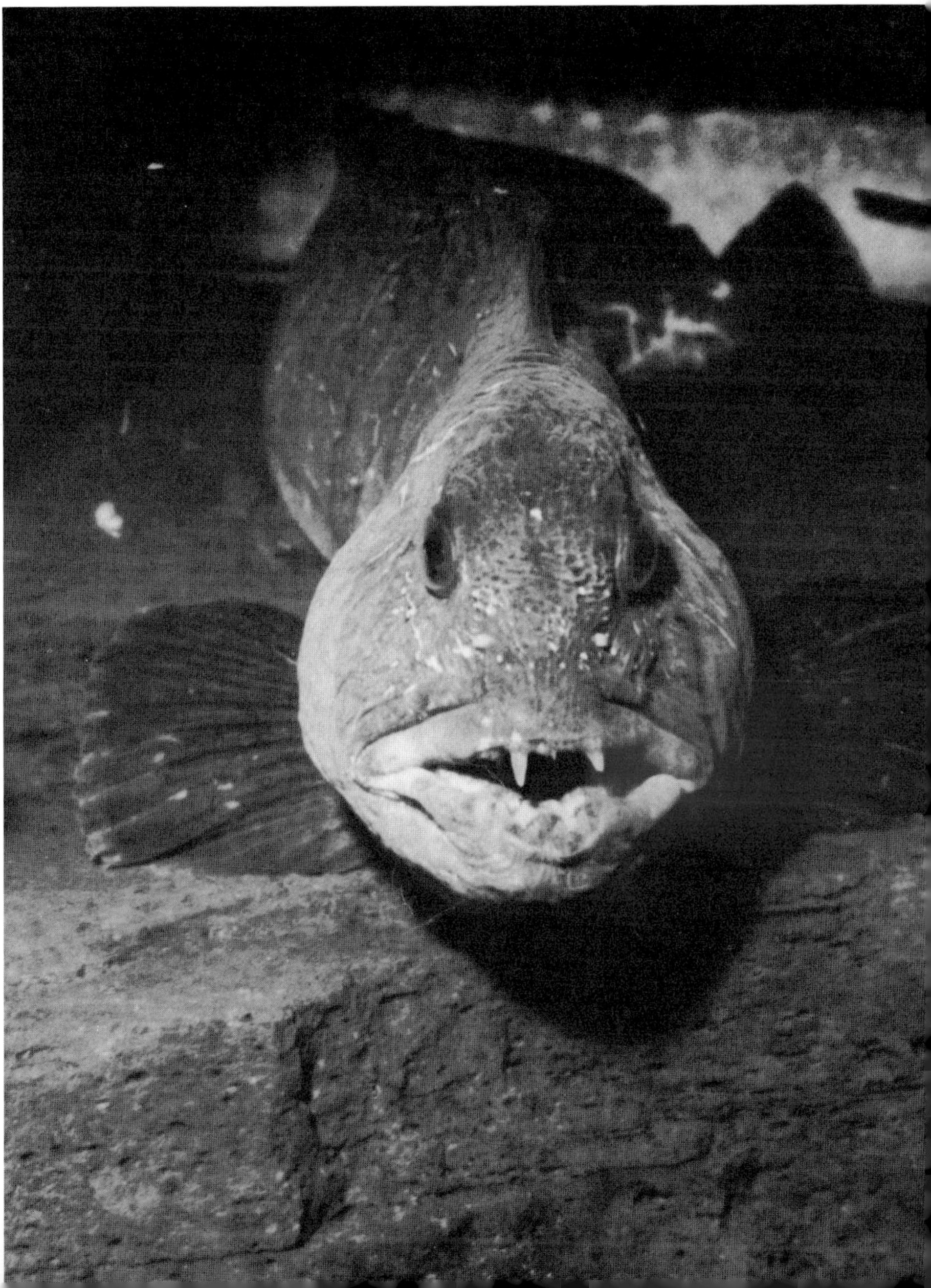

## Axolotl

Few animals seem more misshapen or badly planned than this amphibian. Its gills are external, and it is fish-like, yet legged. It looks and seems like a terrible giant grub. It is, though, an ancient form derived from an ancestor that stood between the fish and the reptile. It is also an animal we do not often see and don't understand well at all. This inoffensive Mexican water creature is as foreign and exotic to us as a creature from another planet. That is strange since it was here to welcome us.

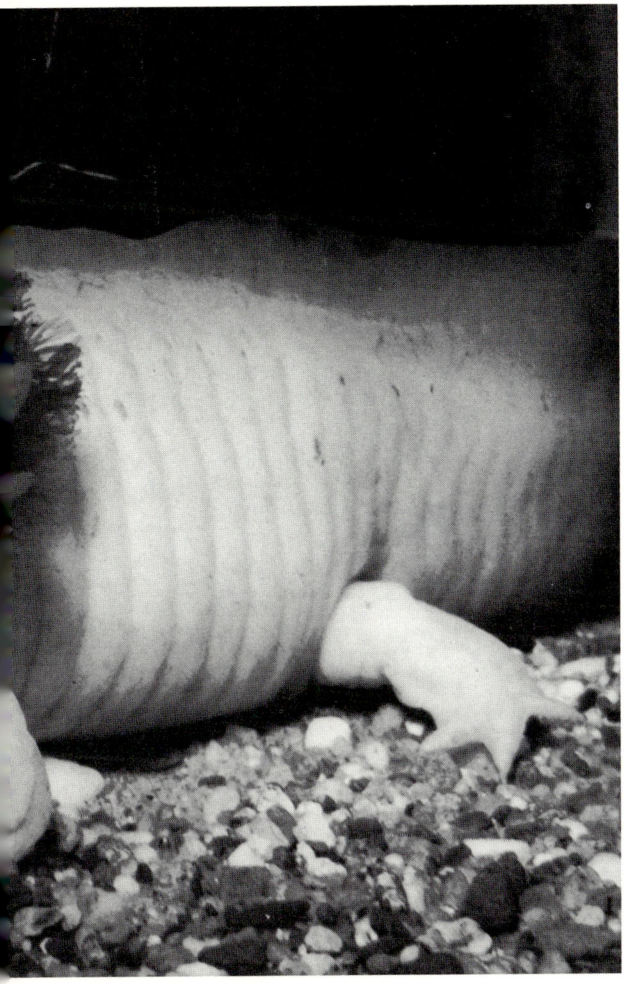

# PHOTO CREDITS

| PAGE | SUBJECT | PHOTOGRAPHER |
|---|---|---|
| 5 | Cicada | Jen & Des Bartlett |
| 6 | Spicebush Butterfly Larva | Carl H. Maslowski |
| 7 | Rhinoceros Beetle | Carl H. Maslowski |
| 9 | Eyed Elator Beetle | Carl H. Maslowski |
| 10 | Stag Beetle | Lawrence Pringle |
| 12 | Moloch Lizard | John R. Brownlie |
| 13 | Ranzania Beetle | Hugo Jelinek |
| 14 | Green Tree Snake | Jane Burton |
| 15 | Box Turtle | Carl H. Maslowski |
| 16 | Chameleon | Jen & Des Bartlett |
| 17 | Marine Iguana | George Holton |
| 18 | Grasshopper | J. H. Carmichael, Jr. |
| 19 | Praying Mantis | Russ Kinne |
| 21 | Rhinoceros | J. W. Cella |
| 22 | Yawning Hamadryus | Tony Angermeyer |
| 23 | Male Stump-tailed Macaque | Jen & Des Bartlett |
| 24 | Proboscis Monkey | Russ Kinne |
| 26 | Young Screech Owl | Richard Frear |
| 28 | Major Mitchell Cockatoo | John R. Brownlie |
| 28 | Ostrich | Russ Kinne |
| 29 | Shoebill Stork | Jen & Des Bartlett |
| 30 | King Vulture | Russ Kinne |
| 32 | Australian Pelican | Eric Lindgren |
| 33 | Saddle-bill Stork | Russ Kinne |
| 34 | Rockhopper Penguin | Russ Kinne |
| 35 | Ant Curtain | Rudolf Freund |
| 36 | Head of Fly | Herman Eisenbeiss |
| 37 | Jumping Spider | J. H. Carmichael, Jr. |
| 37 | Euchromiidae | Fran Hall |
| 38 | Mandrillus Sphinx | Herman Eisenbeiss |
| 39 | Asiatic Horn Frog | Tom McHugh |
| 41 | Chameleon | H. Uible |
| 41 | Frog | Herman Eisenbeiss |
| 42 | Deep Sea Angler | Peter David |
| 44 | Millipede | Robert Dunne |
| 46 | Deep-Sea Squid | Peter David |
| 48 | Yellowfin Toadfish | Tom McHugh |
| 49 | Puffer Fish | Russ Kinne |
| 51 | Platypus | Tom McHugh |
| 52 | Peacock Worm | Russ Kinne |
| 55 | Violet Sea Fan | Russ Kinne |
| 56 | Porcellanid Crab | Jen & Des Bartlett |
| 57 | Zebra Moray Eel | Russ Kinne |
| 59 | Deep-Sea Angler | Peter David |
| 60 | Wolffish | Russ Kinne |
| 62 | Axolotl | Russ Kinne |